Vedic Science and Modern Science

A Difference of Approach

Prof. Ravi Prakash Arya

Amazon Books, USA

In Association with

Indian Foundation for Vedic Science

1051, Sector-1, Rohtak, Haryana, India. Pincode- 124001

Ph. No. 09313033917; 09650183260

email : vedicscience@gmail.com; vedicscience@rediffmail.com

web: https://vedic-sciences.com

First Edition

Kali era : 5122 (c. 2020)
Kalpa era: 1,97,29,49,122
Brahma era: 15,55,21,97,29,49,122

ISBN: 978-81-947593-8-6

Price: 200.00

Printed by

Indian Foundation for Vedic Science, 1051, Sector-1,
Rohtak-124001, Haryana

Contents

Introduction of Sanskrit to the Western World

Introduction of Sanskrit to the west in the last quarter of 18th century attracted western scholarship to the study of Indian literature, history and culture. William Jones (1746-1794) was among those occidental scholars who took the lead in Indic Studies. He founded the Asiatic Society of India in Calcutta in 1786.[1] Two years earlier, he delivered the third annual discourse;[2] in his often-cited "philologer" passage, he noted similarities between Sanskrit, Ancient Greek and Latin as under:

William Jones (146-1794)

> The Sanskrit language, whatever be its antiquity, is of a wonderful structure; more perfect than the Greek, more copious than the Latin, and more exquisitely refined than either, yet bearing to both of them a stronger affinity, both in the roots of verbs and the forms of grammar, than could possibly have been produced by accident; so strong indeed, that no philologer could examine them all

---

1 T. K. John, "*Research and Studies by Western Missionaries and Scholars in Sanskrit Language and Literature*," in the St. Thomas Christian Encyclopaedia of India, Vol. III, Ollur [Trichur] 2010 Ed. George Menachery, pp.79 - 83

2 *A Reader in Nineteenth Century Historical Indo-European Linguistics*: The Third Anniversary Discourse, On the Hindus

> three, without believing them to have sprung from some common source, which, perhaps, no longer exists; there is a similar reason, though not quite so forcible, for supposing that both the Gothic and the Celtic, though blended with a very different idiom, had the same origin with the Sanskrit; and the old Persian might be added to the same family.

Friedrich Max Mueller (1859) calls this event as the beginning of comparative linguistics, Indo-European studies, and Sanskrit philology in the west.

Europeans developed their own subjective bias, dogmas and postulations in the name of 'scientific objectivity' to interpret the culture and history of the colonial people as per their suitability. They developed their own Eurocentric ideology and a historical framework based on their own history which was applied by them and is still applied by their followers to Indian and other cultures inappropriately. This distorted hermeneutic exercise of western scholars presented a flawed view of Indian culture and history. They were obsessed with fitting the data to their postulated and pre-conceived models. In this effort, they first doubted and then completely negated and ignored Indian hermeneutics while interpreting Sanskrit literature, philosophy, culture and history. This resulted in the misleading interpretations of Sanskrit literature, philosophy, culture and Vedas. Here, an humble attempt will be made to assess the viability of the approach of modern science with respect to Vedic Science.

On a keen appraisal of western approach of science it is noticed that by the time of development of western approach towards science, the intellectual scenario of

Europe was dominated by the Church, Emerging trends in science and technology, Evolutionism of Charles Darwin and Karl Marx's dialectical materialism. As such all these three factors became responsible for shaping and designing the western approach towards science. Hereunder, we shall take into account all factors leading to the western approach of science.

Church

European scholarship was suffering from superiority complex due to their political subjugation of the whole world. They were victors and others were victims. They were masters and others were slaves. In master-slave or victor-victim relationship, the victor or master is always superior and victim or slave is always inferior. The victor will look down upon the victim and slave. The victor or master will always expect the victim or the slave to think like him. If the victim or slave will not do so, the victor or master will make him do so. As new masters, they wanted to maintain their supremacy. So, they were trying to understand the culture and history of the colonial people with their own approach, since they knew it well that political subjugation will not last long until and unless they subjugate colonial people ideologically and culturally. It was necessary for them to establish their racial, ideological and historical superiority upon the people of this country. To redeem this goal, they wanted to convert the colonial people into their fold. That is why there was a deliberate attempt on the part of European intelligentsia in the 17th and 18th century to present a distorted picture of Indian history, culture and literature so that an intellectual crisis might be declared in this country. Under the cover of such a crisis they wanted to, as was done by Macaulay, enforce upon Indians their education so that Indians bereft of their cultural richness and superiority in the past might switch over to their faith and pave the way for their permanent settlement and rule in this country. As the part of their plot against Indians, most of the European scholars pursued Indic Studies with the ulterior motive of Christianizing India.

We come across an astonishing fact that Max Müller who himself was deeply interested in ancient Indian texts, served as a functionary for the colonialists and for Christian evangelists. This fact is verified by one of his letters addressed to the Duke of Orgoil, who was the British secretary of state for India, Müller (Georgina Adelaide Müller, 1902: Chp. XVI, p.358) wrote on 16 December 1868:

Max Müller (1823-1900)

'The ancient religion of India is totally doomed and if the Christianity doesn't step in whose fault will it be.'

In a letter addressed to his wife Georgina Adelaide Müller on December 9, 1867, Prof. Max Mueller wrote (Georgina Adelaide Müller, 1902: Chp. XV, p. 328):

'I still have a great work to do, and I often feel that I might have done a great deal more if I had kept the one object of my life more steadily in view. I sometimes wish you would help me more in doing that, and insist on my working harder at the 'Veda' and nothing else. I hope I shall finish that work and feel convinced that though I shall not live to see it, yet this edition of mine and translation of Vedas will hereafter tell to a great extent on the fate of India and on the growth of millions of the souls in this country.'

At the same place, he further observes:

'It is the root of their religion, and to show them what the root is, I feel sure, is the only way of uprooting all that has sprung from it during the last 3,000 years.'

The text of his letter is self-explanatory to the fact that Max Müller pursued Indic studies with an ulterior motive and he was more or less successful in his efforts.

E B Pusey (1800-1882)

His friend Dr. E.B Pusey who was an English churchman and for more than fifty years Regius Professor of Hebrew at Christ Church, Oxford, hurls praises on him for the good job done by him for which he was awarded the position of Sanskrit Chair. He also makes a mention of his support in his elevation to the Chair. He writes on June 2, 1860 (Georgina Adelaide Müller, 1902: Chp. XII, p. 237-38):

'My dear Professor, On the first election to the Sanskrit Chair, you will have heard that we were divided before two great names. Professor Wilson, whose first-rate Sanskrit knowledge was in the mouth of everyone, and Dr. Mill, who, many of us thought, might fulfil the object of the founder better by giving to the Professorship a direct missionary turn. The same thought would naturally recur to us now, and I have kept myself in suspense since our sudden loss of Professor Wilson. **My first impression, however, is my abiding conviction, that we should be best promoting the intentions of the founder by electing yourself, who have already done so much to make us fully acquainted with the religious systems of those whom we wish to win to the Gospel**. It is obvious that without this knowledge a missionary must be continually at fault, ignorant alike of the points of contact of which, after the manner of St. Paul, he may avail himself, or of those which present the chief obstacles to the reception of the Gospel in the minds of those whom he would win. I cannot but think then that your labours on the Vedas-while they attest your wonderful power in mastering this ancient Sanskrit (and of course of the more modern Sanskrit, through which you had access to the older), and while they evince, as **I understand, great philological talent, beyond the knowledge of Sanskrit itself-are the greatest gifts which have been bestowed on those who would win to Christianity the subtle and thoughtful minds of the cultivated Indians**. We owe you very much for the past, and we shall ourselves gain greatly by placing you in a position in which you can give your undivided attention to those labours by which we have already so much profited. You know that I have felt it my duty to confine myself to a different class of languages, those which bear directly upon Hebrew. I have written, therefore, on that upon which I am alone competent to write - not your great knowledge of Sanskrit, of which we have such eminent testimony, but of the great value of that special line of study to which you have devoted yourself. **Your work will form a new era in the efforts for the conversion of India, and Oxford will have reason to be thankful that, by giving you a home, it will have facilitated a work of such primary and lasting importance for the conversion of India, and which, by enabling us to compare that early false religion with the true, illustrates the more than blessedness of what we enjoy.**

Yours very faithfully,
E. B. Pusey.

The above letter by a Churchman to Müller reveals the hidden agenda of colonialists. This is not the single example. There is another celebrated scholar called Monier Williams he writes (1879: 262):

Monier Williams (1819-1899)

'Christianity has many more points to their ancient faith than Islam has, and when the walls of the mighty fortress of Brahmanism are encircled, undermined and finally stormed by the soldiers of the cross, the victory of Christianity must be signal and complete.'

In his preface to his famous Sanskrit-English Dictionary (P ix, 1899) Monier William, as the Professor of Boden Chair, reveals the objective of founding the Chair for Sanskrit studies by Col. Boden as to convert the natives of India to Christianity. He writes thus:

In explanation I must draw attention to the fact that I am only the second occupant of the Boden Chair, and that its founder; Col. Boden, stated most explicitly in his will (dated Aug. 15, 1811) that special object of his munificent bequest was to promote the translation of the scriptures into Sanskrit; so as to enable his countrymen to proceed in the conversion of the natives of India to the Christian Religion.

H H Wilson (1786-1860)

The Book '*Eminent Orientalists: Indian, European and American*' (p. 53) informs that Prof. H.H. Wilson, one of the occupants of Boden Chair, delivered two public lectures at Oxford before general audiences on 'Religious and Philosophical Systems of the Hindus'. These lectures were written to help candidates for a prize of Pound 200 given by John Muir, a well-known old Haileybury man and great Sanskrit scholar, for the best refutation of the Hindu religious systems. The prize was obtained by one Mr. Mullions.

The above evidence shows that the whole exercise was done by Prof. Wilson for 200 Pounds to encourage students for the refutation of Hindu religion.

However, I must say that whole European

community cannot be blamed for their covert designs until and unless some concrete evidence is come across with regard their bias against Indian literature and their objective to evangelize India.

In addition to what has cited above, we must not forget that in the 16th and 17th century, Church was dominating the western intelligentsia. Nobody dared to challenge the ideas subscribed by the Church. We have the example of Galileo Galilei (1564-1642) who had to face the ire of Church when he challenged the geocentrism subscribed by the Church and championed heliocentrism. The entire intellectual world of Europe surrendered before the Church and we find an echo of this fact in a very popular saying, ‘So far as religion is concerned, the sun moves around the earth, but so far as science is concerned, the earth moves around the sun’. Now we may discuss the veracity of the evangelical model of hermeneutics.

Archbishop Usher

(1581-1656)

In the 16th century, an Archbishop named Usher (1581-1656) thoughtlessly declared that creation took place on 22 October, 4004 years before the birth of Christ at 9 AM. Thus a *lakṣmaṇa rekhā* (limit) of chronology was drawn (fixed) and the same was accepted by historians as the borderline in western hermeneutics to fix the chronology of the history of Europe as well as other countries colonized by them. Europeans developed such a framework of their hermeneutics as could not accommodate the concept of history stretching beyond

2000 years before Christ and any chronology beyond 4004 years before Christ was outrightly denounced. Under the pressure and impression of Church, an old chronology of Indian history based on astronomy was debunked and new chronology was assigned demoting it to match with the chronology of the European history.

When William Jones, who started conducting a fair study of the Indian literature, assigned a higher chronology to Indian history on the basis of the records of the Purāṇas than recommended by Church, he was persistently persuaded by Max Müller to demote the Indian chronology to such an extent as could prove the civilization of west far more ancient than that of India. This scheme of Max Mueller of demoting Indian chronology deliberately is exposed by his following request made to William Jones:

> "There is but one means through which the History of India can be connected with Greece and its Chronology be reduced to proper limits. Although we look in vain in the literature of the Brahmaṇas or Buddhists for any allusion to Alexander's conquest, and although it is impossible to identify any of the historical events, related by Alexander's companions, with the historical traditions of India, one name has fortunately been preserved by classical writers who describe the events immediately following Alexander's conquest, to form a connecting link between the history of the East and the West. This is the name of Sandracottus or Sandrocyptus, the Sanskrit Chandragupta.

Max Müller's above-cited insistence forced William

Jones to fall in line with him in misapplying the Greek synchronism of Sandracottus with Chandragupta of Maurya Dynasty instead of the Samudragupt of Gupta Dynasty and thus chronology of Indian history was demoted by 1200 years. Under the same scheme maximum date for the composition of the first book of humanity, the *Ṛgveda*, was assigned c. 1500 B.C. and million years old astronomical chronology assigned by Indians to the origin of the universe, earth and Vedas were mocked and debunked as farfetched one.

Under the pressure and impression of Church, western approach towards Indian history, science and civilization was developed such a way that the sense of history and chronology in India was declared as non-existent. To quote a few, a German scholar M. Winternitz (1927, p. 30) in his Introduction to his reputed work '*A History of Indian literature*' observes as follows:

M. Winternitz (1863-1937)

> 'To them, the facts themselves were more important than their chronological order. They attached no importance at all, especially in the literary matters, to the question of what was earlier or later.'

An Arabian traveller Alberuni (Edward C. Sachau, p. 10), who in the year 1030 wrote a book on India remarks:

'Unfortunately, the Hindus do not pay much attention to the historical order of things. They are very careless in relating the chronological succession of their kings and when they are pressed for information, they are at loss, not knowing what to say, and invariably take to romancing.'

A A Macdonell (1854-1930)

A. A. Macdonell (1900, p. 7) also draws our attention to the same weakness of Indian literature.

'History is one weak spot in Indian literature. It is, in fact, non-existent. The total lack of historical sense is so characteristics, that the whole course of Sanskrit Literature is darkened by the shadow of this object, suffering as it does from an entire absence of exact chronology'.

According to Max Müller (1859, p. 18)

'No wonder that a nation like India cared so little for history?'

While indulging in the Indian history bashing program, these scholars avoided referring to the findings of Jones[3] with regard to Indian history. He observes as under:

> 'And for these generations (Brahaddradhas) the Hindus allotted a period of one thousand years. They reckon exactly the same number (1000 years)

---

[3] Works of William Jones, 13 Vols. chapter on 'The Chronology of the Hindus'

of years for twenty generations of Jarāsandha, whose son Sahadeva was contemporary to Yudhisthira, and founded a new dynasty in Magadha or Bihar. Then there is a list of twenty Magadha kings from Sahadeva, son of Jarasandha to Satyajit, whose son Puranjaya is killed by his minister Sunga who placed his son Pradyota on the throne. Then the Pradyota dynasty starts in 2100 B.C., and then, the Sisunaga Dynasty from 1962 B.C. followed by Nanda Dynasty from 1602 B.C. Then comes the Maurayas from **1502 B.C.**, the Sung dynasty from Pushyamitra 1365 B.C., to Kshema Bhumi 1253 B.C., then the Kanwa Dynasty from 1253 to 908 B.C. followed by the Andhra Dynasty from 908 B.C., to 452 B.C., the last king being Chandrabija".

The above information from William Jones' treatise repudiates westerner's accusations that Indians had no sense of writing history, or chronologically narrating the events. We had clear-cut chronology assigned to each and every event of history. But it was a well-planned attack by the westerners to eradicate India's past. They disapproved of the authenticity of Paurāṇika and other historical records. Had they approved of the authenticity of Paurāṇika and other historical records preserved in Sanskrit literature, they would not have been able to declare the absence of history in India and historical sense of Indians and distort Indian history fact-wise and chronology-wise.

Such western scholars wanted Indians to think of their history in terms of their approach. Whereas, the fact is that in western culture, the concept of chronology evolved 2000 years ago. Since they wanted the supremacy of Church, they accepted the Birth date of Christ as the sheet anchor in the historical chronology of the world. They started teaching in the light of the false Biblical view that Christ was the son of God born first ever on the earth. St. Augustine through Hegel was impressed by Christian fundamentalism and sectarian bias. Karl Jaspers (1963, p.1), the writer of "The origin and goal of history" reveals this fact as:

> 'In the western world, the philosophy of history was founded in the Christian faith. In a grandiose sequence of works ranging from St. Augustine to Hegel, this faith visualised the movement of God through history. God's acts of revelation represent the decisive dividing lines.'

Hegel was proud to say:

> "All history goes toward and comes from Christ. The appearance of the son of God is the axis of world history. Our chronology bears daily witness to this Christian structure of history."

Thus they started calculating the genesis or origin of all events and things after or before the birth of the first son of God, the Christ and vainly claimed themselves as the father of history. Historian philosophers like Jaspers (1963, p. 11) criticizes the idea of history writing based on Christian faith. According to him,

> 'But the Christian faith is the only one faith, not the faith of mankind. This view of universal history, therefore, suffers from the defect that it can only be valid for believing Christians. But even in the west, Christians have not tied their empirical

conceptions of history to their faith. An article of faith is not an article of empirical insight into the real cause of history, as being different in its meaning.'

In fact, our western friends had no idea that the concept of chronology or Kālagaṇanā in India finds its origin with the very origin of the universe. They kept finding in ancient Indian literature the traces of Christian chronological dating type system and so couldn't understand the astronomical dating of the events. Astronomical dating, to them, appeared, as it does, in real sense romances and nothing else. They couldn't make out Kalhaṇa's dating of Mahābhārata war. i.e. Saptarṣi Saṁvat 628. i.e. 3138 BC. To them, Vāyupurāṇa's dating of Nanda's accession in the Saptarṣi Saṁvat 1015 (according to Patna school) i.e. 453 B.C. and in Saṁ. 610 (according to Kashmiri school) i.e. 453 B.C. had no meaning. Varāhamihira's unfolding his own time period i.e., 2625 years after the Yudhiṣṭhira's rule, i.e. 3138-2625=513 BC went unnoticed by them.

In addition to above, postulations of modern science, evolutionism of Charles Darwin, Dialectical Materialism of Karl Marx also influenced the western models of hermeneutics. Those European scholars who remained unfazed by the pressure of their evangelical bosses were impressed by the evolutionism of Darwin or dialectical materialism of Karl Marx.[4] The following analysis will

---

[4] The Marxist theory (adopted as the official philosophy of the Soviet communists) that political and historical events result from the conflict of social forces and are interpretable as a series of contradictions and their solutions. The conflict is seen as caused by material needs.

help us understand the viability of hermeneutics influenced by western science, evolutionism of Darwin and materialism of Marx with respect to Indian literature, culture and history.

Science and Technology

The new emerging trend of science and technology also became one of the factors responsible for shaping the western model of hermeneutics. Hereunder we shall take into consideration some of the prominent issues.

1. Western science is a materialistic science. It studies the matter. Accordingly, the living or non-living world is composed of matter, whereas Vedic science is a spiritualistic science. It considers this creation manifestation of Brahman. As such, there is a basic difference between concept and philosophy between modern science and Vedic science. Any hermeneutic model founded on the materialistic approach will not be able to explain the literature or culture developed upon the pillars of spirituality.

2. Western science tries to understand unknown through known; invisible through visible. Whereas Vedic science has a different approach, it tries to comprehend unknown which is the base of known. If the unknown is realized, the mystery of known will be unfolded itself. In Vedic science, the journey starts from within. The following statement of Chhandogya Upanishad depicts this fact as follows:

कस्मिन्नु भगवो विज्ञाते सर्वमिदं विज्ञातं भवति।

kasminnu bhagavo vijñāte sarvamidaṃ vijñātaṃ bhavati. — *Muṇḍaka Upaniṣad*, 1.1.3

Let me know the reality whose knowledge makes

everything known.

Modern science wants to understand root through branches, whereas the Vedic science studies roots to understand the branches. According to the modern science, branches are above and root below. But according to Vedic science root is above and branches below.

ऊर्ध्वमूलम अधो शाखः ।

ūrdhvamūlama adho śākhaḥ

As such a model of a different approach will fail to explain the system based upon a different approach. Any model dealing with the visible world and branches cannot explain the system of knowledge dealing with invisible world and roots.

3. The Vedic science is a holistic science. It studies things in totality, whereas, western science deals with individual parts. It goes from parts to whole. Nevertheless, it would be very difficult to comprehend the whole based on the parts. It would completely be a conjectural work to determine the shape and structure of a complete body based on parts without having the prior knowledge of the complete body. It will be more or less like the story of four blinds set out in search of an elephant. The blind who caught hold of the elephant by teeth described the elephant as the sword. The one who caught hold of it by tail described it as the broom. Who caught hold of the elephant by legs, described it as a pillar. The one, who caught hold of the elephant by the trunk, described it as a pipe. Thus, they made a puzzle of an elephant. A person who had perfect vision solved their puzzle. He told them that elephant is like a sword,

a pipe, a pillar as well as like that of a broom. All these make a complete elephant. On the other hand, if the whole is comprehensible, the parts will be comprehended automatically. For instance, there will be a total difference of perceptions of the earth, when it is perceived as a whole from the space and when it is perceived from the earth itself in parts. The perceiver of the earth from the space will have a more accurate, precise and quick comprehension regarding the size or appearance of the earth, than the one who tries to construe it from fragments. He may take years and even then, he may fail to draw the right conclusions.

In the other words, the Indian system of knowledge developed based upon the inductive method of inquiry. For the inductive method of inquiry, one has to develop his senses to the extent that he is capable of perceiving the whole. If one wants to comprehend the universal system through the deductive methods, he may have to take innumerable births to reach the ultimate reality, but the inductive method will make one comprehend the whole system at a faster pace. For example, if you develop the habit of reading the page as a whole, your reading will be faster and quicker than the one who reads it word-wise and line-wise. So any methodology based upon a system of analysis of parts cannot be applied to the system of knowledge developed on the cognition of whole.

4. Western science is based on observations. Whatever is observed during experiments in a laboratory is often incorporated in the parameters of science. As far as Vedic Science is concerned, it is both experimental and experiential science. It is not only based on observations; rather it is also based on realisation,

Adi Shankaracharya

(1168 B.C.-477 B.C.)

experience or visualisation. Ādi Saṇkarāchārya (1168 B.C.-1136 B.C.) says ज्ञानं विषय विषयानुभूतिर् विज्ञानम् ॥ *jñānaṁ viṣaya viṣayānubhūtir vijñānam.* That is jñāna is mere information of an object or a thing, but science is the realisation of that particular thing or an object.

Swami Dayanand Saraswati

Maharishi Dayananda Saraswati (1824-1883), one of the leading Vedic scholars and social reformer of the 19th century also defines Vijñāna as तस्य परमेश्वरादारभ्य तृणपर्यन्त पदार्थेषु साक्षाद्-बोधान्वयत्वात्। i.e. domain of Vedic science entails the realisation of all the things right from the grass-root till God. Thus, realisation has far deeper meaning than the observation. Comprehension of Spiritual aspect of the creation is possible through realization or through Aptopadeśa, descriptions of high-profile seers or yogis who could realize the reality. That is why Indian hermeneutics has also included Āptopadeśa as one of the means of knowledge. Vedas and other Śāstras are included into the category of Āptopadeśa. So, it can be said that any method dealing with pratyakṣa means of knowledge cannot be applied to a system of knowledge developed on the basis of realization. In the other words, the knowledge of a mystic can be verified by the mystic only.

5. Approach of western science to provide comforts to human beings, whereas approach of Vedic science was to transforming the life of human beings. So both the systems have a difference approach. A system with a different approach cannot explain the system with a different approach.

Evolutionism of Charles Darwin

Charles Darwin (1809-1882) had emerged a dominating figure in the field of science in 18th century Europe. His thesis on 'The Origin of Species' brought about a revolution in Europe. In his above-cited thesis, he postulated everything evolves gradually. In the beginning, everything remains in the primitive shape and becomes refined by degrees with the passage of time. He emerged as a savant for the scientific community in Europe. His words were taken as the words of Bible. From his idea of evolution and natural selection, Herbert Spencer derived his postulations of 'struggle' and 'survival of the fittest' and Karl Marx derived his postulation of 'dialectical materialism', 'class struggle. These ideas also helped shaped and designed the occidental and Marxist models of hermeneutics which were accepted as scientific methods of inquiry. Everything was assessed on the

Charles Darwin (1809-1882)

touchstone of evolution and struggle.

Since Vedas are the first book in the library of humankind, so they were not taken as the book of knowledge as per Indian tradition, rather they were considered by the western evolutionist scholars to be the literary genre of primitive tribal communities and an attempt was made to glean out the culture and history of tribal people from the Vedas. As per evolutionism, the primitive man got bewildered to have a glimpse of the heavenly bodies and got sacred. So he started worshipping them as gods. Vedic people were also branded as the worshipper of nature. According to Max Müller Vedas cannot advocate monotheism. As the concept of monotheism cannot develop in the early stages of human culture. The mantras depicting monotheism were composed very late. So, primarily Vedas advocated of henotheism. Sri Aurobindo says that Europeans were so much obsessed with the evolutionism that if a Vedic mantra did not support evolutionism, either its meaning was distorted or it was branded as interpolated one.

Similarly based on the postulates of 'struggle', 'survival of fittest', the western scholarship read the meaning of historical warfare in the descriptions of the Vedas pointing allegorically to the natural phenomenon or interaction among the natural forces. For example, Devāsura Saṅgrāma and Indra Vṛtrāsura Yuddha were read as a struggle between Devas and Asuras and Aryans and aboriginals.

Marxism

Marxism is founded on materialism and inspired by Darwin's Evolutionism. From the idea of evolution and natural selection put launched by Darwin, Karl Marx derived his postulation of 'dialectical materialism' and 'class struggle. These ideas of Marx helped shape and design the Marxist model of hermeneutics. I am dealing with this model separately as many of the earliest researchers to venture into Indic Studies were Marxist historians. For instance, from the 1850s on, Karl Marx carefully studied India as a colonial country where diverse forms and methods of colonial rule had been practised. He also took interest in India because she still retained, to a certain degree, relations peculiar to primitive communal society (Marx, 2001). He also studied the freedom struggle of India (Marx, 1986). The first major Indian scholar to apply the Marxist theories to Indic Studies was D. D. Kosambi (Kosambi, 1956). Let us examine how Marxist scholars have applied their hermeneutics to Indic studies.

Karl Marx (1818-1883)

Marxism is rooted in materialism, whereas, Indic tradition is rooted in spiritualism and dharma. Spirituality and dharma are a fundamental ideology of every sphere of Indian life, sociology, culture, politics, economics and even family life. Therefore, any scholar based solely on Marxism can hardly study a tradition rooted in spirituality and dharma. Marx was an atheist, a rationalist and he had no knowledge of Indian culture and philosophy. His knowledge of India based on the descriptive elements in Hegel's interpretation of the Indian civilization. Hegel saw in the ideology of the Hindu culture, a pantheism of 'Imagination' expressed in the universal deification of all finite existence and degradation of the Divine, a deprivation of man of personality and freedom. Interestingly, Hegel himself had a flawed understanding of Indian culture and philosophy. He had only a limited number of authoritative sources at his disposal; and he was not a person with an open mind, capable of sympathizing with all alien ways of thinking with loving understanding, but was an armchair scholar, inclined to abstractions, interpreting the outside world according to his pre-conceived pattern. Therefore, whatever he was able to say about India turned out to be extremely inadequate, resulting in a distorted picture. Although his portrayal of India had some well-observed details, it shows that he had ventured upon a task, for the fulfilment of which he had not possessed the pre-requisites (Glasenapp, 1973).

Marx opposed religion because his experience with Semitic religions made him believe that religion was contrasting to science and that it hindered the scientific progress. According to him, 'history of religion is the history of the fight against the development of scientific thought.' This charge may be true of medieval Christianity but the case is entirely different when it comes to the Indian context. He was not aware that Indic tradition never professed the type of religion he experienced. Indic tradition professed dharma which may be described as a value system beneficial for the material advancement and spiritual upliftment of a human being. Dharma also denotes duties of an individual towards his family , society and nation. Dharma doesn't hinder science but promote a scientific point of view. In India, dharma and science have always co-existed under one generic name Vidyā, with a subtle distinction being drawn between 'Parā Vidyā', spiritual knowledge and 'Aparā Vidyā', material knowledge. In fact, Marx studied religion from books which he interpreted according to his understanding of them and applied his analysis of history to them. He did not have any knowledge of Sanskrit language and hence could not read the original texts but had to depend on others' interpretations. While the Indic dharma is based on direct realization, Marx did not have any such experience or even any contact with anybody having such experience. Based on Judeo-Christian notions, Marx considered religion to be born out of one's helplessness and fear. He considered religion to be a myth promising a better life, later on, the alleged reward for sufferings on the earth. In contrast, Indic dharma is not based on any fear or helplessness. Hinduism teaches that happiness and immortality are to be realized within one's own itself, not to be found through religion. According to Radhakrishnan, Marxism has become a faith with its uncritical supporters (Radhakrishnan, 1948).

Without adequate knowledge of Sanskrit and other Indian languages, Marx and his followers misinterpreted important Vedic traditions. For instance, Yajña was defined in the following way. 'Ya' means to gather, 'J' means to take birth and 'Na' is the suffix added to it. Therefore, Yajña is defined as to live together in the tribes and to procreate together in a community. Also, Brahma is interpreted to mean commune. However,

Raimundo Panikkar (1980) has criticized such translation attempts and has opined that only the person who really speaks the language, that is, who is fulfilled in it, can be a genuine translator.

Shripad Amrit Dange (1949) quotes a Sanskrit verse from the *Aitareya Brāhmaṇa*

> कलिः शयानो भवति संजिहानस्तु द्वापरः ।
> तिष्ठंस्त्रेता भवति कृतं सम्पद्यते चरन् ॥
> तस्मात् चरैवेति चरैवेति ॥
>
> *kaliḥ śayāno bhavati saṁjihānastu dvāparaḥ* ।
> *tiṣṭhaṁstretā bhavati kṛtaṁ sampadyate charan* ॥
> *tasmāt charaiveti charaiveti* ॥

and interprets it to suit his Marxist model in the following way to explain the chronology of ancient times. On the basis of the word '*charan*' associated with Kṛtayuga, he conjectures that in the Vedic period, in Satyayuga, people were in the nomadic stage. Later in the Tretāyuga of Vedas, they settled in tribes and started community farming, tribal procreation without marriage institution. Here he made the word '*tiṣṭhan*' associated with Tretā as the basis of his conjecture. After that in the Dvāparayuga, community farming developed into the agriculture based on possessive rights. Land owners hired slaves and exploited them. In this conjecture, he made the word '*sanjihāna*' associated with Dvāpara as the basis. The last era is Kali Yuga which has the most developed machines, industrial development and hence will bring the ultimate comfort and prosperity. The word '*śayana*' associated with Kali was made the ground for this conjecture. Thus, preconceived hermeneutics based upon postulations of western scholars were applied for interpreting the ancient Indian history and culture giving rise to its distorted portrayal.

Similarly, the sexual relationship is also interpreted in Eurocentric chronology. A statement from Bhīṣma from Mahābhārata is quoted to show that the man-woman relation developed in the following chronology. 'Saṅkalpa' is completely random sexual relations among men and women in the Satya Yuga. 'Sansparśa' is sexual relations except among the same caste men/women in the 'Tretāyuga' of Vedas. 'Maithun' is periodical marriages in the Dvāparayuga. 'Dvandva' is modern marriage in which women are given more conservative role but men are free from any restrictions. Men treat women slavishly in the Kali Yuga. People used to live in the tribes without any notions of family. Human relations such as brother-sister, husband-wife were absent and the only form of relation was mother-child thus portraying the scientific Indian marriage system wrongly which finds its sanction from the Vedic times. The above view of marriage holds good about the western and European people who could not develop marriage system and the human relations until recently. Instead of marrying each other, they used to live in 'live in relationship' like wild animals, as they considered marriage as the hurdle in their physical pleasure. All their relations like the father in law, mother in law, brother in law did not develop naturally as in India but are based upon legal sanctions. The history tells that in Europe till 2nd century AD people were not allowed to marry by the Emperors. During the reign of Claudius Gothicus in Rome, it was St. Valentine, a Roman priest who went against the edict of the emperor and helped young couples in marrying secretly. Valentine was arrested by

Claudius and prosecuted upon being caught marrying Christian couples. On the other hand, in India since Vedic times 10 types of marriages are prescribed. We have different names for the relations like mama (mother's brother), chacha (father's younger brother), tau (father's elder brother), mausa (mother's sister's husband) etc. but all these relations are addressed by one word in the west and that is 'uncle' if male or 'aunt' if female. It shows the poverty of language and relations. Benedetto Croce (1966) mentions that religion, ethics, culture, and marriage institution are products of the disciplined society of modern period and were absent in the ancient India of the Vedic period. Surprisingly he completely discarded the true facts that ethics, culture and marriage intuitions were developed to the highest level ever in the Vedic age itself.

Benedetto Croce (1966) in his book 'Historical Materialism' mentions that the origin and development of everything is based on materialistic need. Based on this theory, Marxist scholars have deliberately attempted to label every ancient tradition into the categories of materialism. Therefore, the social institutions of India like Varṇa system is seen through the lenses of class war and exploitation of lower castes by the higher castes. Although the historical evidence has proved that the present caste system is the product of British colonization of India in the last couple of centuries (Dirks, 2001). Similarly, spiritual philosophers such as Shankara are also alleged to misguide the lower class of people with the ideas of devotion and faith to appease the higher class. The Marxist approach has tried to establish the socio-economic factor as a mantra to explain every facet of Indian history and culture. The fact that Marxists had become dogmatic and mechanical in their analysis was criticized by D.D. Kosambi (1956), who is regarded as the pioneer among Indian Marxists. Their application of dialectical materialism to the interpretation of Indian culture is grossly mechanical rather than critically discerning (Goyal 2000).

According to the Marxist scholars, Indian history is just a series of successive invasions. But even a cursory study of the several periods proves this observation to be superficial. Another Marxist myth is that Asiatic societies have been unchangeable. However, even from the materialistic perspective, Indian society has had massive changes. For instance, the distinction between the stage of food-gatherers and that of food-producers stands completely shaken by the excavations at Kalibangan where evidence has become available of plough-farming having been known in India as early as the times of the Saraswati Vedic civilization. Similarly, India does not seem to have ever had a classical slave economy in the same sense as Greece and Rome had. Indian economy never depended on large-scale chattel slavery, which according to orthodox Marxism inevitably produced the feudal stage of history. And whatever loose kind of feudalism there may have existed in India, it cannot be said to conform to the orthodox Marxist definition. Similarly, the Varṇas in ancient India cannot be identified with the economic classes of Marxism. Indeed, the concept of social class, which is a key component of Marxism, has invited many theoretical objections, and its application to all periods of history has not always proved fruitful. In fact, there is no adequate evidence available of any conscious class struggle in India (Goyal, 2000).

Noted Sanskrit scholar Daniel H. H. Ingalls (1965) had also identified the flawed Marxist hermeneutics applied by Marxist scholar D.D. Kosambi. Ingalls rejected Kosambi's application of the Engels' and Plekhanov's theories of the class origins to Indian art and literature. In the six to seven centuries of India's history from the end of the Gupta dynasty to the coming of Islamic invaders, there arose no new social class of note. In obedience to the theory of Engels, there should not have been any significant art during this period. But those are precisely the greatest periods of art, not only in poetry but in painting, sculpture, architecture and philosophy. Therefore, any foreign hermeneutics cannot be applied to none of the ancient traditions with a subjective bias. Ingalls concludes that the path to a proper understanding of Sanskrit literature must begin with Sanskrit literature itself. In finding one's way one must seek guidance from the scholars versed in the tradition who can understand and interpret it. If one is finally to condemn any tradition it must not be done by a foreign theory which Indian authors never professed, but by the principles of mood and suggestion which they claimed to follow. After one has in mind clearly what the texts mean and what their authors were seeking to achieve one may go on to compare the general principles of Sanskrit texts with those set forth by Aristotle etc. It is at that point that one may take into account the differences in social structure between ancient India, Greece and modern Europe. Those differences might have played an important role in bringing about the variation of character among different kinds of literature. But the path of the critique of texts must begin with the texts, not with the theories of society.

In view of the aforesaid discussion, it can be concluded without any hesitation that with particular reference to ancient Indian history and culture, all different aspects of human life - religious, political, social, economic and cultural - should be studied as different facets of one organic reality, each of them studied in its relation to others and not in isolation from each other. Moreover, all human institutions cannot be forced to follow the same line of development everywhere. The laws of social change are by no means as predictable as the laws of physics. The western approach full of prejudices and preconceived notion will present Indian history and culture in a wrong perspective. For instance, it would tend to eclipse or misrepresent what has been generally regarded as India's peculiar glory and her particular contribution, namely her richly varied religious philosophies. It follows from the above discussions that Indic Studies can be pursued in proper perspective with the help of Indian hermeneutics.

Role of Science & Technology

Modern age is the age of Science. Science has played the vital role in the making of modern society. Modern science has contributed to the making of modern society by way of developing the technological known how. In the other words, the main thrust of modern science has been towards the *sanskāra* (refinement) of material things. The modern technique of refinement, modification, purification and processing the material things is known in Vedic science as *sanskāra*. Thus we can say that modern science has given stress upon the *sanskāra* of material things and developed a technology to promote the convenience and comfort of human beings. Here there is a basic difference in the concept of Vedic Science and modern science. Vedic science gave more thrust on the *sanskāra* of human beings as compare to material things. According to the Vedic science, a human being who has not gone through the certain process of *sanskāras* remains in his crude form and is recognized as no better than an animal. As such a human being is not considered to be the actual social being. In material things also we can see, so long as they are in their crude form, they cannot be used practically. For their practical use, the material things are to undergo a particular *sanskāra* or processing. For instance, an oil in its crude form cannot be used, but after undergoing the process of refinement it becomes usable. Thus the conversion of oil from its crude form to the refined form is known as Science. Similarly processing of other material things for the use of human-beings is known as Science. In the race for developing technology modern Science has totally ignored the need for *sanskāra* of humans. That is why, today in the age of machinery, human beings are also considered to be as machines. Today the human beings are considered as life less things and so there is always a talk about human resource development never as human development. The concept of human resource development lies in considering humans as resources just like other natural resources. Natural resources are often exploited for the benefit of humankind and so are human beings. However, there is basic difference in the exploitation of natural sources and exploitation of human resources. Exploitation of natural resources may benefit the entire humankind, on the other hand exploitation of human resources is not always going to benefit the entire humankind. Exploitation of human resources turns usually in the exploitation of certain helpless human beings for the larger interest of certain opulent group of human-beings.

So far as Vedas are concerned the prime duty of science is not only the *sanskāra* of material things, but the *sanskāra* of human beings also. Vedic seers never considered human beings as the means or resources. For them the human beings are not the means but ends. Every thing was directed towards the development and elevation of human beings.

Vedic seers had the humanitarian and ethical approach. Their main aim was to the relieve the human beings permanently of the suffering, deprivation and all other problems that haunted them.

Modern Science has though developed so many aids for the comfort of human beings, yet it has a limited approach. Its approach is not Universal. Modern science has no moral, ethical or humanitarian foundation. It is founded on the laws of survival of fittest in the struggle that ensues among human beings on the globe. Modern science wants to define and describe everything in the light of struggle. It has no remedy to put an end to this struggle. Contrary to this, main aim of Vedic science is to put an end to this struggle and prepare a stage for friendship, fraternity and co-existence of each and every being on the Earth. It clears a stage for co-existence of each and every human-being by giving a call for accepting the principals of mutual understanding and co-operation among human beings.

सह नौ भुनक्तु सह वीर्यम् करवावहै ।
तेजस्विनावधितमस्तु मा विद्विषावहै ॥

saha nau bhunaktu saha vīryam karavāvahai,
tejasvināvadhitamastu mā vidviṣāvahai.

"Let there be mutual co-operation in eating and gaining power. Let each one of us become illumined with knowledge. let us not envy or struggle with each other."

Since the modern science has its foundation laid out on the laws of struggle its talk about the exploitation of natural as well as human resources aims at the comfort of those who are able to survive the struggle on the globe. We can see that in spite of high advancement in the area of technology and production of each and everything, the benefits are not reaching to each and every human being on the globe due solely to the absence of moral and humanitarian values. We have more than sufficient amount of food grains piled up in our stocks, still there is poverty and hunger on the globe. At one hand, the excess food grains are being disposed off in the sea, on the other hand human-beings and other animate beings are dying of hunger. The Science has blessed the modern humanity with great prosperity still more than 50% of human beings on the globe are below poverty line and suffering from hunger, thrust, malnutrition, phenomenon of coldness and hotness and dearth of essential things for their survival. One can say unhesitatingly that modern Science has failed in performing the *sanskāra* of human beings. It could develop and produce things but not real human beings.

Everything has been commercialized. In the race of

commercialization moral values have left far behind. Money has become the basis of material advancement but moral values are the real basis of the advancement of humanity on the globe. These moral values are called by *Vedic* seers as *Dharma. Dhāraṇāt Dharma ityāhuḥ* i.e. since these values are sustaining human life on the globe, they are called Dharma. Following moral values are said to be the constituents of *dharma.*

Dhṛti (patience), kṣmā (forgiveness), damaḥ (to have control over one's mind), asteya (non stealing), śauca (inner and outer cleanliness), indrīya nigraha (to have control over one's sensual desires) dhi (use of intellect, rationale) vidyā (education) satya (seeking truth) and krodha (calmness, i.e absence of anger). Today we see Dharma is applied in altered sense. Different sects or schools propagating and fanning fundamentalism are known as religion, which is not dharma in factual sense.

The *sanśkāra* of human beings takes place by inculcating the above components of Dharma (i.e. moral values of life)

There is clear cut observation in the Vedic philosophy that every human being is born as śudra from the womb of his/her mother. After undergoing the process of *sanśkāra*, he becomes dvija (twice born).

जन्मना जायते शूद्रः संस्कारैर्द्विज उच्च्यते ॥

janmanā jāyate śūdraḥ saṁskārairdvija uccyate.

That is a human-being born physically is not better than an animal. It is the *sanśkāra* (inculcation of moral values and education) that makes him a real human being. For becoming the actual human being, he has to take second birth, i.e. in addition to his physical birth from the mother's womb, he has to take birth form Āchārya's womb. Āchārya is a person who educate him and help inculcation of moral and humanitarian values.

अचार्य कस्मात् आचिनोति अर्थान् आचारम् ग्राहयतीति सतः ॥

āchārya kasmāt ācinoti arthān ācāram grāhayatīti sataḥ.

Just like a child stay for nine months in his/her mother's womb, similarly he has to stay with Āchārya in Gurukula for years together to be born again as a perfect, knowledgeable socially fit human-being. The stay of child with Āchārya is called as *antaḥvāsa* and the child observing *antaḥvāsa* (staying in the womb of Āchārya) is known as *antevāsī.* Just as in developing a certain instrument or machine, several types of processing is required, similarly in developing a perfect human being, the Vedic sages stressed the need of sixteen sanśkāras right from the birth till death of an individual. Modern Science's advancement is lop-sided. It has played pivotal role towards developing perfect machinery and has cared less in the development of perfect human beings.

So far we have confined ourselves to the physical and mental *sanśkāra* of the human beings. For maintaining the physical health, researches are going on, but for the upkeep and maintenance of spiritual health or character building no research or effort has been put in. Whereas Vedic scientist was equally concerned about the spiritual health, as was with the physical and mental health.

Vedic ideas are still very much relevant for the upkeep and advancement of modern society. The need of hour even today is the moral and spiritual education, inculcation of ethical, moral and humanitarian values and spiritual advancement, in addition to the modern formal education so that the modern society may rid of vices of struggle, exploitation, commercialisation of each and every thing and may be endowed with moral / humanitarian and spiritual values. This is how, all human beings may transform into perfect human beings (Āryas) and remove hunger, poverty, depravedness from the globe through their mutual co-operation instead of being compartmentalised and divided along the lines of cast, race, region or religion. Let the whole world sing the Vedic song.

भूमि माता पुत्रोऽहं पृथिव्याः

Bhūmi mātā putro'ahaṁ pṛthivyāḥ

"Earth is my mother and I am the child of mother earth."

REFERENCES

Croce, Benedetto. (1966). *Historical Materialism and the Economics of Karl Marx*. New York. Russell & Russell.

Dange, Shripad Amrit. (1949). *India from Primitive Communism to Slavery*. Bombay. People's Pub. House.

Dirks, Nicholas B. (2001). *Castes of Mind: Colonialism and the Making of Modern India.* Princeton, N.J. Princeton University Press.

Eminent Orientlists: Indian, European and American'. (2000). Cosmo Publications India.

Edward C. Sachau (1910). *Alberuni's India*, 2nd edition. Kegan Paul, Tubner and Co., London.

Georgina Adelaide Mñller. (1902). Ed. *Life and Letters of Frederick Max Mñller*, Vol.1, Longmans, Green, And Co.

London.

Glasenapp, Helmuth von. (1973). *Image of India.* New Dehli. Indian Council for Cultural Relations.

Goyal, Shankar. (2000). *Marxist Interpretation of Ancient Indian History.* Post-graduate & Research Department series, no. 43. Pune: Bhandarkar Oriental Research Institute.

Ingalls, Daniel H.H. (1965). *On the Passing of Judgments, an Anthology of Sanskrit Court Poetry, Harvard Oriental Series,* Vol. 44.

Jaspers, Karl (1963). *The origin and goal of history,* New Haven and London, Yale University Press.

Kosambi, D. D. (1956). *An Introduction to the Study of Indian History.* Bombay: Popular Book Depot.

Macdonell, A.A. (1900). *History of Sanskrit Literature,* D Appleton and Company, New York

Marx, Karl. (1967). *The Wisdom of Karl Marx.* New York: Philosophical Library.

________ (1986). *Notes on Indian History* (664-1858). Moscow. Progress Publishers.

Max Mueller, F. (1859). *History of Ancient Sanskrit Literature, Williams and Norgate, London*

Monier Williams. (1879). Modern India and Indians.

Panikkar, Raimundo. (1980). 'Hermeneutics of comparative religion', in *Journal of Dharma,* (V, 1), Jan-Mar 1980.

Radhakrishnan, S. (1948). *Religion and Society.* London. George Allen & Unwin.

William Jones. (1799). *The Works of William Jones,* G.G. and J. Robinson, London.

Winternitz, M (1927) '*A History of Indian literature,* Vol. 1'. University of Calcutta.